THE TRUE STORY OF THE 1899 SHIPWRECK OF THE "SCOTSMAN"

At Belle Isle between Newfoundland and Labrador

PAUL LANGAN

HISTORICAL LANGUAGE ADVISORY

Certain parts of this book contain historical language and content that some may consider offensive, for example, language used to refer to racial, ethnic, and cultural groups. Items in the book, their content and their descriptions reflect the time period when they were created.

The items retain their original descriptions to ensure that attitudes and viewpoints are not erased from the historical record.

ISBN 978-1-998829-00-2

The True Story of the 1899 Shipwreck of the Scotsman -
At Belle Isle between Newfoundland and Labrador

Compiled by Paul Langan

brentunespublishing@yahoo.ca

paullangan.com

CONTENTS

INTRODUCTION

The story of the shipwreck and subsequent tribulations of the passengers and crew of the "Scotsman" in 1899 is a quintessential part of our Canadian maritime history. Sadly, it is being forgotten as time passes. The purpose of this book is to give you the most complete information on this event that will be preserved for generations to come.

On September 22nd, 1899, just 4 years after it was built, the Dominion Line "Scotsman" ocean liner ran aground at Belle Isle. Belle Isle is a small uninhabited Island off the northern tip of Newfoundland. It is between Newfoundland and Labrador.

The horrendous ordeal passengers and crew had to experience are well documented in this book. It contains newspaper articles and eye-witness accounts from survivors. One of the survivors had taken photos which are included in this book. There was an inquiry under the Wrecks and Salvage Act into the action by the captain of the ship. The report and decision are also included in this book.

NOTE: There are various numbers mentioned in the book about how many people died. This is due to the date of the various references used. The actual number is 24. Some drowned when the first lifeboat was lowered into the water. Others perished due to the hardships encountered on Belle Isle.

This story of survival would make for an excellent movie. Enjoy reading this important part of our Canadian history.

Paul Langan

Hespeler, Ontario

November 3rd, 2022

DOMINION LINE AND SCOTSMAN HISTORY

The Scotsman Ocean Liner - credit - www.wrecksite.eu

The Dominion Line Royal Mail Steamers was involved in the transatlantic immigrant passenger service between Liverpool, Queenstown (Cobh) and Boston, Liverpool to Quebec and Montreal and Naples to Boston. They were in operation under the Dominion Line from the late 1800s to 1908, and White Star Dominion Line until 1926.

These Steamers are fitted with Electric Light throughout, have Saloons and State Rooms, Music Rooms, Smoking Rooms and Bathrooms amidships, where the least motion is felt, are amongst the largest and finest afloat and are well and favorably known for the comfort and excellent arrangements they afford for all classes of Passengers.

SS Scotsman

One of the largest steamers in the Canadian trans-atlantic service being over six thousand tons. On the fatal voyage, the vessel and cargo easily represent one and a half million dollars.

Her passenger arrangements were very fine, and her saloon was handsome; finished in bird's eye maple and satin wood.

General

Nationality: British

Purpose: Transport

Type: Ocean liner

Propulsion: Steam

Date built:1894.

Details

Tonnage:7101 grt

Dimensions:143.5 x 15 x 9.7 m

Material: Steel

Engine: 2 x 3 cyl. triple expansion steam engines, dual shaft, 2 screws, 4 masts

Power:499

Speed:13 knots

Yard no.:289

IMO/Off. No.:105283

Boilers: She has four boilers, two of which were double ended, each 17 feet in length and all four tested to stand n pressure of 180 pounds to the square inch.

About the loss

Cause lost:

Ran aground (wrecked), Fog

Date lost:22/09/1899.

Casualties:

† max.13 rank: 724

About people

Captain: Skrimshire, Henry Truscott

Crew:120

Passengers:279

Owner

Dominion Line - Liverpool & Mississippi S.S. Co. - North Atlantic Steam Nav. Co., Liverpool

Builder: Harland & Wolff Ltd., Belfast

Engine by: Harland & Wolff Ltd., Belfast

About the wreck

Position

Latitude: 51°55,569'N

Longitude: 55°25,233'W

System: WGS84

Reliability: Bad >1km

Remarks: Calculated as '7.44-mile S W [225°] from Belle Isle N.

Lighthouse Last update: Jan Lettens, 31/12/2010

History

SCOTSMAN SS was 6041 gross tons, length 470.7ft x beam 49.2ft, one funnel, four masts, twin screw, speed 13 knots, accommodation for 100-1st, 130-2nd and 800-3rd class passengers.

Built by Harland & Wolff, Belfast she was launched for the Dominion Line on 13th December 1894 and started her maiden voyage from Liverpool to Halifax and Portland on 28th November 1895.

On 14th May 1896 she commenced her first Liverpool - Quebec - Montreal voyage. On the 22nd September 1899 she was wrecked in the Belle Isle Straits with the loss of 13 lives.

About Owners - About Builders

Harland & Wolff Ltd.

Harland and Wolff were formed in 1861 by Edward James Harland and Hamburg- born Gustav Wilhelm Wolff. Based in Belfast at Queen's Island, Harland and Wolff are a huge and very important shipbuilding company. The shipbuilding complex is only one of two yards left in the U.K. capable of building large merchant ships. The yard was most well known for building high-class transatlantic passenger liners and was considered to be the best in the world. The company has built over 1700 ships at four yards and has been in operation for over 135 years.

Dominion Line - Liverpool & Mississippi S.S. Co. - North Atlantic Steam Nav. Co.

Dominion Line is the short name of the Liverpool & Mississippi Steamship Co., founded 1870 and renamed Mississippi & Dominion Steamship Co. in 1872 two years later. She was absorbed by International Mercantile Marine Co. (Leyland Line) in 1921

Harland and Wolff Shipyard, Belfast Queen's Island, Ireland - credit wrecksite.eu

References - https://www.wrecksite.eu/wreck.aspx?139956, http://www.theshipslist.com, http://www.gracesguide.co.uk, https://www.ggarchives.com/OceanTravel/SteamshipLines/DominionLine.html#gsc.tab=0

DOMINION LINE
LIVERPOOL
TO CANADA

WHERE IS BELLE ISLE?

Belle Isle (French for "Beautiful Island") is an uninhabited island slightly more than 24 km (15 mi) off the coast of Labrador and slightly less than 32 km (20 mi) north of Newfoundland at the Atlantic entrance to the Strait of Belle Isle, which takes its name.

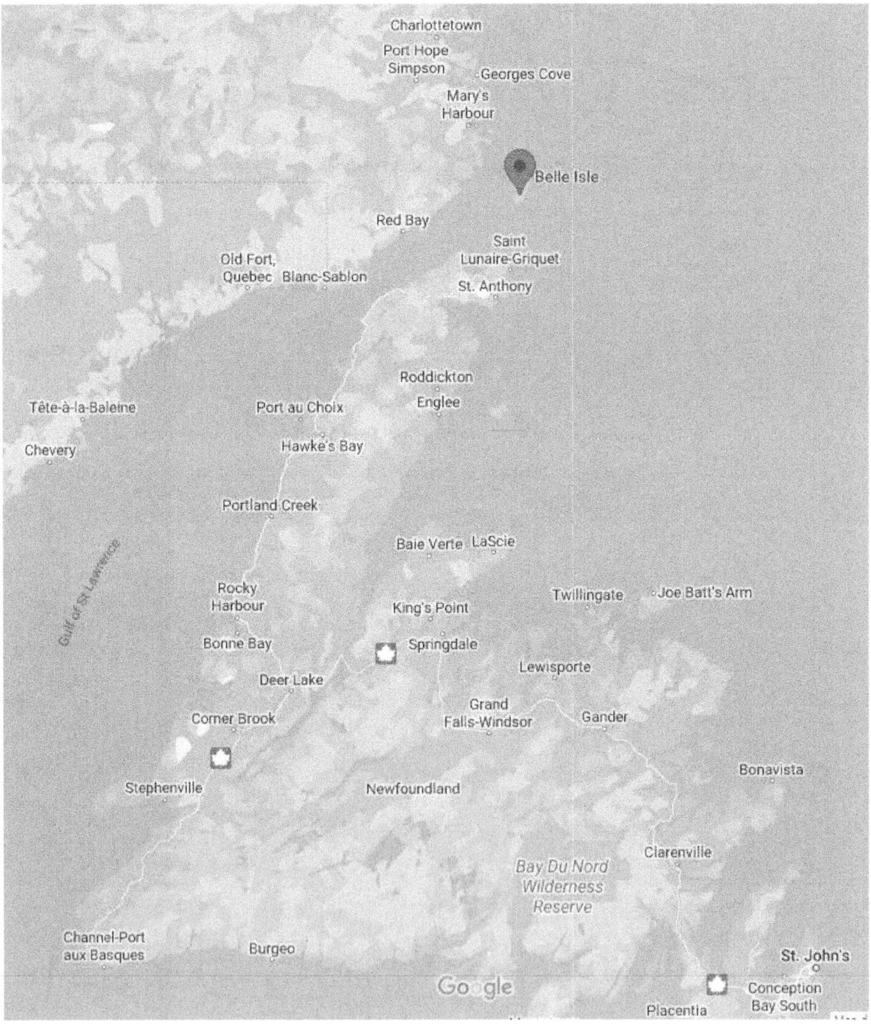

TEN PERISHED.

The Details of the Wreck of the Scotsman.

INTENSE SUFFERING

Passengers Four Days on Rocks With Little Food.

ROBBED BY THE FIREMEN.

"Scabs" Shipped at Liverpool Went Through the Passengers' Baggage — Survivors Arrived in Montreal This Morning.

Quebec, September 29. —(Special)— The first authentic news of the wrecking of the Dominion line steamship Scotsman was obtained this morning when the passengers who had been picked up by the Elder, Dempster steamship Montfort, arrived at Rimouski on their way to Montreal.

The story told by the passengers of the wreck, and their subsequent experiences until they were rescued off the coast of Belle Isle by the steamship Montfort, is most sensational. The Scotsman, which left Liverpool on September 14th, bound for Montreal, went on the rocks in a dense fog on the southeast coast of Change Island, eight miles southeast of Belle Isle lighthouse, at 2.40 on the morning of Friday, September 23rd. The passengers were all below in their berths at the time and were awakened suddenly from their sleep by the signal foghorn, followed a few minutes later by a terrible crash as the ship went on the rocks. The passengers rushed on deck, men, women and children, with hardly a stitch of clothing on them, to find the ship rocking to and fro on the rocks.

The night was pitch dark, there was a high wind, which amounted almost to a gale blowing, and the sea was running high. Captain Skrimshire, who remained on the Scotsman's bridge, fearing that the vessel would topple over and sink with all on board, at once gave orders to have the boats lowered, and issued peremptory instructions that the women were to be attended to first.

Capt. Skrimshire and his chief officers displayed great coolness, and the calmness of the passengers under the circumstances was something remarkable. There was no panic, and very little confusion; and the passengers, with a few exceptions, showed great sang-froid in obeying the officers' instructions. Five boats were launched, and the women were lowered into them. The launching of the first boat resulted in the loss of at least ten lives, and it is feared that a number of others perished. The boat which was lowered with the women sank almost as soon as it touched the water, and its thirty occupants were left struggling for their lives in the icy water. Some of the sailors, who were in charge of the other boats, behaved heroically, and succeeded in rescuing a number of the drowning passengers, but many sank to rise no more.

List of the Dead

The passengers who were drowned by the sinking of the boat so far as can be ascertained were:

Saloon Passengers

Mrs. Robinson, Miss Robinson.

Mrs. Child, wife of Harry Child.

Miss Street, Montreal, travelling companion of Miss Duncan, a wealthy Montreal lady.

Second Cabin Passengers

Mrs. Dickinson, of Windsor, Ont., who was returning from a visit to England.

Miss R. Weavers, of London, Eng., who was on her way to Montreal.

Mrs. Talbot, London, Eng.

Mrs. Scott, of Barnet, London, Eng., who was on her way to Port Arthur.

Miss. Watkins, England.

Miss Harrison.

Infant child of Mrs. Roberts, England.

It is supposed that the sinking of the boat was caused by one of the plug holes being open, which allowed the boat to quickly fill with water. Some of the other boats narrowly escaped a similar mishap, and disaster was only prevented by those in charge stuffing something into the plug holes in order to prevent the boats from filling. The occupants of the four remaining boats, for the most part frail and delicate women, were tossed about on the angry sea near the wreck for ten or eleven hours without food or provisions of any kind. The majority of them were scantily clad and suffered terribly from the cold and exposure.

Early on Friday morning the fog cleared a little, and the vessel was still holding fast, the captain ordered the boats back to the vessel. In the meantime, as the fog had lifted with the breaking of day, it was found that the rocks on which the vessel had run formed part of the huge mass of cliff on the Belle Isle coast, and a number of the male passengers were lowered by means of ladders to a ledge of rock. When the boats returned to the vessel the women were hauled over the vessel by means of ropes and lowered to the ledge in chairs and by means of the ladders.

The passengers were now safely off the rock on solid earth, but the worst of their troubles were to come. It was necessary to get out of the danger of the incoming tide and to do this they had to climb the steep cliff, a height of a couple of hundred feet, to reach a place of safety on a plateau above. It was not so bad for the men as for the unfortunate women, many of whom were in their bare feet and with hardly anything but their night dresses on then.

Robbers at Work

To add to the gravity of the situation the firemen and lower grade sailors, who were simply "scabs" taken on at Liverpool owing to the strike, behaved in an atrocious manner. No sooner had the vessel gone on the rocks than they started to loot the ship from end to end. They entered the passengers' cabins, broke open the luggage and stole everything they could lay their hands on. They broke into the steward's quarters, broached the liquors, and filled themselves with wine and whiskey until they were in a beastly state of intoxication. To the appeals of the women who asked for provisions and assistance they replied with jeers and drunken oaths.

While the women were almost perishing with cold on the barren plateau these monsters in human form stalked about the rocks drinking and singing obscene songs. The conduct of Capt. Skrimshire and his chief officers all through was most praiseworthy, but the lower grades of the crew, and especially the firemen and stokers, set all authority at defiance and their conduct was a disgrace to humanity.

The three hundred passengers passed all day Friday and Friday flight on the narrow plateau a couple of hundred feet above where the vessel was wrecked. The night was a beautiful moonlight one, and the scene's the passengers declared were most weird and solemn. A number of the passengers sang hymns and Father Langmore, a Church of England clergyman, offered a prayer for their safety. A party which had been sent out by the captain to hunt for the lighthouse returned early on Saturday morning and reported that the lighthouse was about twelve miles distant, but the journey to it was something terrible.

During Saturday, Sunday and Monday, the passengers in groups covered the distance between the rocks and the lighthouse. At the most it was only fifteen to eighteen miles, but to nearly all it seemed thousands. The way over the steep cliffs, and through swamps and waste places which had never been trodden by the foot of man before. There were innumerable ponds and brooks through which the footsore passengers had to walk. But all kept a stout heart, and though a number of them were exhausted by the way, the majority reached the lighthouse in safety.

For four days, from early Friday morning until Monday night, they had wandered on the rocks with hardly any food, and with only the water they could obtain from the brooks to drink. The women, though the most fragile, were amongst the bravest and struggled on nobly united they reached the lighthouse, when many fainted away from exhaustion. The passengers were most hospitably treated by the lighthouse keeper, Mr. Coulton and his sister, who, although very short of provisions, did all they could for them.

On Tuesday morning the Elder- Dempster steamship Montfort, which was passing Inward to Montreal, was signalled from the lighthouse. The Montfort at once responded to the signals, and all the passengers, except about fifty, who had not reached the lighthouse, were taken on board, and brought to Rimouski.

It is expected that the passengers who were left behind will be picked up by the Allan line steamship Grecian, which is on her way up, and be brought to Quebec.

Capt. Skrimshire, and a number of the crew are standing by the vessel, which it is feared may be a total loss.

As to the cause of the disaster the following story is told: —On Friday morning, while in a dense fog a whistle was heard. Captain Skrimshire says, according to the story, that he concluded it was that of another steamer and altered his course accordingly. A few moments later there was an awful crash, and the vessel was upon the rocks. Formerly the fog signal on the end of the island was a gun, but a short time ago a whistle was substituted, and this was the whistle that was heard. The captain, it is reported, asserts that he was never notified of the change and so it was that the disaster occurred.

Reference - 10 Perished, The Montreal Gazette, 1899/09/30 page1

FRANK H. WESTERON, PASSENGER AND PHOTOGRAPHER ON THE SCOTSMAN

We read about shipwrecks every day, but it is to be feared they make little impression on the majority of us, mainly because we cannot realize the horrors of these dread tragedies of the sea, unless we have actually experienced them. We venture to think, however, that Mr. Westerton's set of new and unique snapshots will convey an entirely new impression of a shipwreck, showing as they do almost every phase of a terrible experience. The author is a well-known member of Mr. William Greet's "Sign of the Cross" Company.

We left Liverpool on the evening of Thursday, September 14th last, in the fine Dominion liner *Scotsman,* having a full complement of passengers and crew. In consequence of the then prevailing strike of sailors and firemen the vessel did not lie alongside the landing- stage, so that we had to be taken out to her by tender. The first two or three days passed in the usual routine of life on board a liner; the only thing we had to vary the monotony being some rather heavy weather, which was marked by a very strong head wind, increasing gradually to a gale, and rendering our progress very slow.

My first snapshot, taken from the deck, gives you a fair idea of the kind of seas we encountered. People have asked me whether I had any presentiment of coming evil at the moment when I snapped this little photo. My reply is that I had not. It was my intention to take photos, of most phases of life on board, and of the landing of passengers when we reached our destination. Little did I dream, indeed, what the rest of my set of snapshots were going to be like. We were so much hampered by the rough weather that one day's run was well under two hundred miles for twenty-four hours. On Thursday, 21st of September, the sea was much calmer, but there was no sun, and presently it began to get a little hazy, with a slight suspicion of fog. We were all particularly happy that evening, however, and broke up into merry whist-parties in the saloon and smoking-room. At this time everyone was looking eagerly forward to seeing land the following morning. We *did* see land; but Lord! under what ghastly circumstances!

A fair idea of the kind of seas we encountered when Mr. Westerton took this photo. He little dreamed what the rest of the set would look like.

Most of us were rather later than usual in retiring that night. As far as I can remember it was half - past two o'clock on Friday morning, the 22nd of September, when we were all awakened by a fearful crash, followed by a series of terrible bumps and grinding gratings. In an instant vague voices arose, and each and every soul on board felt an awful sensation of terror and sickening foreboding. My first impression was that we had struck an iceberg, and I said so to my saloon companion, who immediately afterwards hurried on deck to see for himself what had happened. After a few minutes he returned to say that we were on the rocks. He also suggested my dressing immediately—a most unnecessary admonition, considering that I was tearing on my clothes as fast as I could. A minute or two later everyone on board appeared to be astir, and the passages were crowded with men and women with blanched faces and curiously varied expressions of terror.

The situation of the ship at this moment is exactly as you see it in the second snapshot, except, of course, that the ladders seen on the left were not run out, nor were the passengers swarming over the rocks as you see them in the photo. Simultaneously with, the crash, every light on board went out for a few seconds, rendering the situation appalling in the extreme— what with the incessant grinding of the unfortunate ship on the cruel rocks; the piercing whistle of the bo'sun; the deafening siren; the escape of steam from the boilers; and the confused shouts of crew and passengers.

As you may suppose, it took me but a very few seconds to make my way to the upper deck, and then the full meaning of our position began to dawn upon me. The fog was very dense indeed, and we could hardly tell whether we were at sea or close to the shore. The unfortunate *Scotsman* had taken a terrible list to port, and it seemed as though she must founder every minute. Men, women, and children were running about in a meaningless, erratic sort of way, clad for the most part in weird nightdresses. Some of them were fastening on lifebelts more or less calmly. I noticed that the men did this chiefly, at the same time assuring the women that there was no danger. The captain (poor fellow, I pitied him 1) was on the bridge shouting orders through the megaphone, but what with the escape of steam and the terrible excitement I fear his voice was scarcely heard. And yet, miraculous though it may seem, you could not say there was a panic, the womenfolk especially behaving with marvellous pluck and obeying the commands of the officers. This absence of panic was, I feel sure, chiefly due to one of the engineers, who, as we afterwards learnt, waded through the water in the engine-room to the starboard side, his object being to keep the electric light going as long as possible.

After a few minutes (of course it seemed hours to us) the port boats were lowered, and above the uproar came tire captain's order: "Women and children into the boats." About seventeen people were lowered into the first boat, which was promptly swamped by the heavy sea then running. I leave the scene that followed to your own imagination. Everything had been going well and the excitement was rapidly subsiding, when we saw these poor creatures drowning before our eyes, shrieking, and screaming for help in the most agonizing way. With all possible speed the next boat was lowered, and every endeavour humanly possible was made to save the drowning people. I grieve to say, however, that only three or four got safely into the second boat. By the way, it required the greatest care to keep the other boats from being swamped likewise. This was due not only to the heavy sea running, but also to the fact that the huge vessel lurched with every wave that came, in a way that hurled men and women onto the deck or against the taffrail, causing serious falls and injuries.

Although everything was done that pluck and good seamanship could suggest, it was some considerable time before all the available boats were filled with passengers, and finally the captain ordered them to stand out about a mile from the shore until daylight came. By this time the fog had lifted slightly, and we found ourselves within a few yards of a terrible coast. Every effort was now directed to finding the best possible means of landing the remainder of the passengers and crew.

Several ladders were lashed together and thrust out of the starboard gangway, as you see in the photo, just previously reproduced. A sailor then clambered along the ladders, tied a rope to one of the end rungs, and let himself down onto the rocks. This done, he hauled the ladder ashore and made it fast. Then the work of rescue, which it took several hours to complete before everybody was ashore. Of course, the captain was the last to leave the vessel. His conduct was magnificent — quite in accordance with the finest traditions of British seamanship. This was about half past six; and you must remember that the boats, with their human freight, were still some little distance out at sea.

This snapshot shows at a glance the position of the ill-fated liner cast upon that terrible coast.

The passengers had to climb these precipitous rocks to reach their miserable camping ground.

After a consultation the captain ordered everybody to make for the higher rocks in case the ship foundered, or the tide should rise. I can assure you that it was quite a wonderful sight to see old men and women and young children scaling the precipitous rocks as shown in the above illustration. The ship's papers had been rescued, besides several sacks of bread; so that when we came to review our position, we shook hands and congratulated one another on our miraculous escape from what had at one time appeared certain death. It was now evident that the *Scotsman* was stuck firmly on a ledge of rock and was tightly wedged there. Only a few yards away there was a great depth of water, so that had she struck but a little way astern she must have foundered at once.

It was not until about noon on Friday that the boats, which had remained out at sea, were able to come in and land the women. This was done by hauling each person up again over the ship's side, and then down on to the rocks by means of the ladders. My next photo shows the rocks as seen from the ship, with thc people climbing up towards the camping-ground above. The women who had been in the boats were in a truly pitiable condition, many of them wet through and through, and clothed for the most part merely in night-dresses and the lightest of wraps. It was a terrible time, too, for those who were eagerly watching and waiting to know whether their friends and relatives in the boats were or were not safe; and when all were landed the suffering and grief of those who had lost wives, mothers, or sisters was heartbreaking to witness. Altogether it was found that fourteen people had been drowned by the swamping of the first boat.

The doomed "Scotsman" as seen from the camping ground 200ft above.

As the captain had thought, we had struck on the coast of Belle Isle, and after a kind of council of war, men were sent out to try and find the lighthouse which was known to exist not very far away. They returned next day, however, without having found it, but on a second voyage of discovery the welcome structure was found to be about ten miles to the south. When all the people were safely brought up to a kind of plateau about 200ft. up the rocks, we pitched our camp there, although on the first day we had only dry bread and water, of which latter commodity there was a plentiful supply running down the rocks. Besides, later we found there were many large fresh-water lakes scattered over the island. The snapshot reproduced here will give you an admirable idea of what the unfortunate *Scotsman* looked like from the camping-ground 200ft. above sea-level.

The rocks as seen from the ship with passengers on their way up to the camp.

The unfortunate passengers huddling together in miscellaneous garb and varying degrees of misery.

That first night we all lay huddled up together to keep each other warm, but notwithstanding this we passed a night of most ghastly wretchedness and misery. I remember distinctly that it was a brilliant moonlight night, and rockets and signals of distress were sent up at intervals, soaring and hissing in a melancholy fashion over our heads, and followed by eager eyes in the hope that they would speedily bring relief. An admirable idea of the "camp" may be gained from the above photo, which shows the unfortunate passengers huddling together in miscellaneous garb and varying degrees of misery.

Morning dawned with no ship in sight. During the second day two vessels did pass in the distance, but our signals were either not noticed or were disregarded. I hardly think the latter was the case, however. Another of my snapshots shows the passengers with the captain and his officers on the look-out for passing vessels. They are all discussing ways and means, for at such times everyone thinks his own plan is the best. After waiting some time, the captain suggested that everyone who felt equal to the exertion should walk over to the lighthouse, where they would find food and shelter. Moreover, there would, he said, be a far greater likelihood of being taken off by passing vessels. A number of passengers at once started, but the journey across the island was found to be a very formidable undertaking indeed. The moss was very deep and boggy, and the ground terribly uneven, rendering the journey one of very great hardship to people wholly unaccustomed to such violent exercise. It was only possible to walk very slowly, and the strongest men among us took six or seven hours to reach the lighthouse, ten miles away. Several of the bravest of the women were compelled to lie down and sleep for the night half-way across, and their sufferings were terrible to witness.

The captain, his officers, and some passengers were on the look-out for passing vessels.

"OUR COMMISSARIAT"— tinned meats, Biscuits, and other provisions were brought from the wreck.

Tinned meats, biscuits, and other provisions were brought from the wreck in the most approved Robinson Crusoe fashion, and in this way, we were kept from starving. My next snapshot is a view of our commissariat, with a specially appointed official attending to the food. Yet another view of our headquarters shows the doctor of the *Scotsman* making soda-water. Several other members of the crew and the stewards are also seen in this little photo.

On Sunday morning a vessel answered our signal and stood in to help us. She was the ss. *Monterey,* of the Allan Line, bound from Montreal to England. Owing to her having the mail on board, however, she would only take off those going back to England. Five or six passengers went on board, besides the majority of the sailors and firemen. We were heartily glad to see the last of these gentry, for on the very night of the wreck, as must be well known to readers of the English newspapers, our staterooms were plundered, and valuables and clothing of every kind stolen by these scoundrels. They were, I am glad to say, placed under arrest on board the *Monterey,* and when they landed in England, they found the authorities prepared for them. As a matter of fact, they were searched the moment they stepped ashore, and those with stolen property on them were promptly dealt with by the law.

On Monday, the majority of those who were left on the rocks started for the lighthouse. Some of these "explorers" endured awful experiences. Many had to sleep on the damp bog half-way across, because they were not able to reach the lighthouse before night fell. I could tell a great deal, did space permit, of the many gallant deeds performed by passengers in helping the weak and infirm. It turned out afterwards that one or two persons had died owing to the fatigue and exposure on this journey, for walking was painfully difficult even for the strongest. In some cases, those who left the "camp" at ten o'clock on Monday morning did not arrive at the lighthouse until twelve o'clock noon on Tuesday! And yet the distance was a bare ten miles!

The ship's doctor is making soda-water for his charges.

The last I ever saw of the *Scotsman* is shown in the snapshot here reproduced. On Tuesday morning the steamer *Montfort,* of the Elder Dempster Line, answered our signals at the lighthouse, and sent her boats to take off the castaways. She took off in all about two hundred and fifty persons, and then from the lighthouse went on to the wreck to rescue the invalids and those who were left behind; also, the captain and several of the officers of the ill- fated liner. Before leaving for Montreal the *Montfort* signalled for assistance to the Allan liner *Grecian.* Captain Jones, of the *Montfort,* as well as his gallant officers and crew, did all they could to make us comfortable and happy; they, as well as the few passengers they had on board, gave up their berths to the ladies of our party. On Friday morning we landed at Rimouski, in the Province of Quebec, and were taken on to Montreal by special train, arriving there at two o'clock on Saturday morning.

As to the *Scotsman,* I learnt afterwards that she disappeared during a heavy storm on Sunday, October 8th last, and now lies buried many fathoms deep, as she would infallibly have done on that fateful 22nd of September but for the merciful Providence that watched over us all.

The author's last glimpse of the unfortunate liner.

Reference: A Shipwreck in Snapshots. The Dominion Liner Scotsman on the coast of Belle Isle. An uncommon original article from the Wide World Magazine, 1900, by Frank H. Westerton

MISS LILLIAN BOYLE AND MISS MARY WILKINS DESCRIBE THEIR EXPERIENCES

Vivid Stories of the Scotsman Wreck Told by Young Lady Passengers on Their Arrival in Vancouver—How They Faced Death for Three Cold, Dreary Days.

Two lady passengers, who arrived from the wrecked Dominion line steamship Scotsman on Monday, gave a vivid description of their exciting experiences to a reporter of The Province. In relating the stories, they were often visibly affected with the memories of the struggle against what appeared sure death, and the hardships of the three long cold and dreary days they spent. They tell many pathetic incidents while They tell many pathetic incidents which will be new and most interesting to the reading public.

Miss Lillian Boyd from Liverpool, Eng., who, with her younger sister Ella, was on her way to Portland, Ore., told a story which corroborated all that Mr. Hoult had said in the interview published in yesterday's Province, with the exception that she was inclined to a more generous view of the chief officer's conduct in not furnishing a better stock of provisions for the long rough trip across the rocks to the lighthouse. "I think," she said, "the captain underestimated the distance and difficulties. He promised to send a boat with supplies to the lighthouse, although for some reason it never turned up."

Speaking of the exciting scenes on the ship Miss Boyd said: "We were roused from our sleep about half past two in the morning by an awful crash and then everything was dark, and we could hear screams and cries and rushing of feet in all directions. Every moment we expected to be our last. We experienced all the sensations of those who face death. My sister Ella is only 35 years old. and although terrified, as I confess, I was, she acted bravely. We dressed as hastily as we could in the darkness, but my sister could not find her boots. Then we stumbled to the main deck.

What we saw and heard there made us forget our own danger. The mother had lost her child —the child the mother. Husbands and wives and friends were parted the same way. Some were praying, some cursing, while hysterical shrieks of men and women in their night clothes and without boots glided here and there like so many ghosts and this was all we could see in the blackness of the night. But it would be impossible for me to give you a conception of the scenes or our feelings. We saw the first boat lowered, and later saw it dashed to pieces and then we heard the last dying screams of its unfortunate occupants. This renewed and increased our suspended terror. Now death appeared to us only a question of time. I wonder now whether we ever ventured in the second boat, which we did. My sister and I were lowered to our boat, which was a large one.

But before I go so far with my story. I might say something about the conduct of the miserable stokers and sailors. I could never have believed that human beings could be so devoid of human sympathy. It would be a reflection on the dumb animal to call these men brutes. Would not a young mothers' cries for her helpless child be expected to move even the heart of stone? But they were nearly all drunk and may not have been so responsible as I thought them at the time. They were entirely out of the captain's control. The male passengers all acted bravely. Mr. Hoult, who is with us on the train, and many others like him, worked hard and were most untiring. Their first object was the safety of the women and children. Every effort was made by the officers and active male passengers to save the wrecked boat's passengers, but so far as I could see, only two or three were saved.

The boat into which we were lowered was a large one and contained thirty women and two children. All the men they could get to row us was one sailor and two stewards. But they worked very, very hard and got us clear of the ship and rocks. I think their intention was to take us to the lighthouse. However, the sea was so rough that they could make no progress. We clung to each other expecting every wave would break over us. Fortunately, we did not get far from the ship. As the dreary, cold hours passed the terrors of death seemed to diminish. And then the dawn began to break, and the waves seemed to grow less. The tired oarsmen renewed their energy and seeing the ship was still above water they thought it best to return to It. But this was no easy matter. We seemed to be always in a round hollow of water and at the mercy of the waves. We were twelve hours on the boat.

in the dreariest cold, and terrifying hours that a human being could experience, when the men by their superhuman efforts again brought the boat alongside the ship. The men on the ship had been watching us and were ready with ropes. They pulled us up one by one, placing the ropes under our arms. By this time, of course, we could see the rocks plainly, and we did not wait on the ship but immediately passed over it to the other side and were then lowered to the ground. We had a big climb to reach the top of the rocks. They said the rock was 400 feet high. Many of the women had no boots on and their bare foot were bleeding at every step. My sister's feet were all cut and are yet much swollen, as you may see. All they gave us to eat for our journey was a sea biscuit each and a cup of coffee between four or five of us.

We were joined by others for our trip across the rocks. There were now 55 in the party. Here the scenes were frightful again. At places on the rocky road the moss was very deep, and we would sink to considerable depth. The weaker women were continually foiling from exhaustion and some fainted. The weakest of our party, Mrs. Bates, was continually delaying us by fainting. At last death released her from her miseries and we left her, having no doubt she was dead.

Others of the women had to be carried the last part of the Journey. They became utterly helpless and would beg us to spare them further torture and save ourselves. We must have travel led 25 miles over these cutting rocks before we reached the lighthouse: Altogether we were over two days on the rocks and the cold, miserable time and the depressing scenes we witnessed I shall never forget. The great kindness shown us by the officers and sailors of the Montfort could not be exaggerated. They submitted to privation themselves for our comfort."

Miss Wilkins' Story

Miss Mary Wilkins, from Cowbridge, South Wales, on her way to Nanaimo, also described similar exciting scenes to the above.

"After the crash," she said, "I rose and dressed myself as well as possible in the darkness, but 1 could not find my boots. In my excitement, expecting death every minute, I rushed to the deck to learn the worst. But all was dark and damp. There appeared to be a heavy fog. The ship was shaking violently, and I expected it to go over. We were all ordered down to a deck below the saloon, and after about an hour some candles were found and lit.

I shall not attempt to describe the scenes. Many were praying; children were crying, and many women were hysterically screaming while the oaths of the drunken stokers and sailors could be heard as they carelessly staggered about. The officers seemed to have no control over them. I saw the first boat lowered but refused to go in it. A personal friend of mine, whose acquaintance I had made on the ship, went down in the boat, and, I think, drowned.

A Most Pathetic Scene

I witnessed the lose of the baby boy of Mrs. Roberts, of Montreal. He was 18 months old and a heavy baby. Mrs. Roberts held him in her arms, and she sat on the boat side. In his terror he struggled. She lost her hold of him, and he fell into the sea and was seen no more. The mother was distracted with grief. I refused to get into any of the boats, because I thought myself safer on the ship, especially after seeing the fate of the first boat.

"The sailors and stokers acted in a brutal manner to the passengers and refused to give any assistance. In some cases, I heard threats of violence made. A Mr. Scott, who could not find his wife and was distracted, was told to 'shut up' by some of the men or they would throw him over the cliffs and then he would be able to join her. His only offence was his repeated appeals, which should have moved the hardest heart.

Bawdy Songs

"Rev. Father Langmore. who was on board, held divine service on the rocks Saturday and Sunday, but our most ardent devotions were rudely marred by the bawdy common songs of the men, who appeared to be entirely without feeling. Our party, which consisted of 15 women and two men and a guide, started for the lighthouse. The guide did not know the way, and we were 17 hours on the road. Many of the women of our company often

Dropped from exhaustion and begged to leave them to save themselves. Mrs. Emerson, of Calgary, had a girl of one and a half years old to carry. I and others helped her as well as we could, but it was very hard work even to walk.

At last, the guide found the path to the lighthouse. Two or three of us then went on and left the rest to follow, and we just reached the lighthouse 15 minutes before the Montfort started for Rimouski. We were glad to reach it. The Montfort had a crew of kind-hearted sailors. They did not have much food on board, but they went short so as to give us plenty. It is under. such conditions that one can appreciate a kindness such as this was.

"All the 18 of our party had to eat on our 17 hours' trip across the rocks was a tin of corned beef, weighing perhaps about six pounds. I don't know why they would not give us biscuits or more provisions. I must say that the stewards acted in a very mean way and requests for food were often answered insultingly. I have no complaint against the upper officers of the Scotsman."

Miss Wilkins was met at the C. P. R. depot by friends, with whom she is now happy at Nanaimo, after her exciting experiences.

Reference - The Brutal Pirates, 1899/10/06 The Province, Vancouver BC page 6

The "Scotsman" Belle Isle - credit britishhomechild.com/the-scotsman

ARRESTS ARE MADE

ARRESTED WITH VALUABLES IN THEIR POSSESSION.

Arrests are Made of the Pillagers of the Wrecked Scotsman and Jewellery Recovered

Twenty-Four Men Taken into Custody-Some Released, Others Retained

Four thousand dollars' worth of jewellery, clothing, and other property, supposed to have been stolen from the passengers of the ill-fated steamship Scotsman, which was wrecked oft Belle Isle, nearly two weeks ago, was recovered yesterday morning from twenty-four persons, who arrived in Montreal shortly after 7 o'clock on the steamship Montfort. The men who gave their occupations as sailors, stewards, and stokers, were placed under arrest by Sergeant Campeau, acting chief of the city detective department, and after being searched at headquarters and the Salvation Army Lighthouse, all but five were allowed to go.

The names of those who were held during the day are: George Parry, sailor; Liverpool; Edward Sweeney, sailor, Ireland; Samuel Wrigley, steward, Liverpool; John Gaade, steward, Birkenhead; and William Bailes, steward, Liverpool. The first and last named were unable to secure bail and were held to await the action of the court.

The others who were arrested by the detectives, but subsequently released, are: Ernest New, Alexander Cormack, William Patterson, John Edwards, W. J. Peters, R. Buntin, D. U. Owens, Thomas Tuhay, Fred Collens, Harvey Kcrhel. William

Colless, Albert Bailey, John Gwythes, Edward Foracres, John Puplett, Angus Morrison, P. Gough and W. H. Williams. Silks, opera glasses and various other valuables were found in their bags. However, the officers decided not to hold them at the City Hall at least, not for the time being, but it is understood that the men are under strict police surveillance.

How Arrests Are Made

Before the gangway had been thrown out a lot of men got their feet in the open portholes and were over the rails before the officers of the vessel could prevent them.

From the alacrity of the officers, it might be judged that they expected what was coming and used excellent judgement in the way they repulsed people they were not quite sure of.

Sergeant Campeau drew his men up, and before there was a chance for any of those having property not belonging to them to dispose of it, they were on their way to Central station in the patrol waggons. At headquarters the bags were searched, and the foregoing property was recovered.

The men are charged with stealing from the steamship Scotsman. They were arrested without warrants on suspicion of larceny, the complaints being made by Mr. Michael Scanlon, superintendent of the Dominion line. Judge Lafontaine said that he could not Issue the regular writs until he had positive information and details. However, the court instructed the detective department that if any goods and property was found, which they suspected had been stolen, the officers would be justified in making the arrests.

One Taken from Grecian

Though there were eleven men taken from the steamship Grecian which arrived in port at 2 o'clock, W. Hamich, a horseman, was the only one placed under arrest. Among his possessions the officers found a horseshoe brooch and a valuable ring.

While the arrest of the twenty-four men from the Scotsman was quiet and orderly there were disturbing elements in evidence when the men from the Grecian were marched to the station. The arrests were made by the detectives, and a number of officers in plain clothes. As they marched the prisoners to the station the crowd followed. Some made ugly remarks, while others talked of lynching, but the people who seemed the least concerned were those in custody of the officers.

Plenty of Plunder

Sealskin sacques, rings, watches and Chains and scores of pieces of costly wearing apparel were found in the sea bags of most of the men who were escorted to the station by Mr. Campeau's men. Only one of the men was able to explain how he secured possession of the stuff. He was John Gaade. When the steamship Scotsman struck the rocks at Belle Isle a child of Mr. Roberts drowned, and he desired to save what little belongings he could. Gaade, who was among the crew, told Mr. Roberts that he would see what could be done, and made special efforts to carry out his promise. He found a number of things In Mr. Roberts' cabin and brought them to Montreal with him. They were found in his bag, and Mr. Roberts yesterday identified them. He also told his story at police headquarters and Gaade was released. He is now the guest of Mr. Roberts.

Deny Charges of Theft

Despite the incriminating evidence piled up against a number of the people arrested on the Montfort not a few of the twenty-four suspects deny the charges that were indirectly made against them, one of them saying that the stuff in the cabins was not touched until it was found it would be a dead loss. As for the stories of looting and the robbery of the contents of the passengers' cabins, they deny all personal knowledge, saying that most of that was probably done by the stokers, who were searched at Quebec.

Arrangements are now being made by the authorities to sort the property for the purpose of identification. The parties who are supposed to be the owners will be notified as soon as possible, after which specific charges will be laid. The Crown has taken up the preparation and it is not at all unlikely that the steamship company will assist in pushing the various cases.

Reference - Arrests are Made, The Gazette, 1899/10/02, page 5

Dominion Line Timetable -
timetableimages.com/maritime/images

WRECKS AND SALVAGE ACT REPORT AND COURT DECISION

(Case No, 5963 -" SCOTSMAN" (S.S.). Canada. Wrecks and Salvage Act

In the matter of a formal investigation held in the Vice-Admiralty Court, Quebec, on the 23rd, 24th, 25th and 26th days of October, 1899, before W. H. Smith, R.N.R., Wreck Commissioner, assisted by Bloomfield Douglas, R.N.R., Naval Assistant, and M. P. McElhinney, Nautical Adviser to the Marine Department, as Nautical Assessors, into the cause of the stranding and total loss of the British steamship " Scotsman," on the N.E. part of the island known as Belle Isle, at the entrance of the strait of the same name, between Newfoundland and the Labrador coast, on the morning of the 22nd of September, 1899.

Report and Decision of Court

The Court having heard and carefully considered the evidence given by the master and the officers and other members of the crew in the master's presence, and attentively listened to the addresses of the counsel employed upon this case, finds as follows:

That the master, Henry T. Skrimshire, gave the ascertained position of the ship, at noon of the 21st, as in lat. 53° 24' N., and long. 50° 04' W., which was obtained by the two morning observations, worked by double altitudes, and reduced to noon by dead reckoning. The lighthouse on Belle Isle was then supposed to be 216 miles distant. At noon the course was N. 87° W. by compass (S. 64° W. true), and the vessel was proceeding at a speed of 13 knots per hour, the weather being fine, with an overcast sky and southerly wind.

That at 6.45 the course was altered to N. 89° W., and at 9 p.m. to S. 86° W. by compass (S. 59° W. true). That the weather remained fine but overcast until midnight, with smooth water, the speed of the ship being estimated to be from 13 to 13.2 knots per hour.

That the said master, in pursuance of his duties, took personal charge on the bridge at 1.46 a.m. of the 22nd. The wind was still moderate with slight mist at intervals, as shown by the logbook. He remained on the bridge until the vessel was stranded.

That at 2.10 a.m. the weather became foggy, the ship still proceeding on the same course, S. 86° W., and going about 13.2 knots per hour; the master then reduced the speed of the ship to 6 or 7 knots, at the same time ringing the engineer on the "Stand by" in the engine-room.

Shortly after this the fog began to thicken, and at 2.20 a.m. it was a dense fog, and there is ample testimony to show that objects could not be seen much further off than the length of the ship. That no soundings were taken at any time to test the accuracy of the assumed position of the ship.

That about 2.25 a.m., dense fog still prevailing, and the night being very dark, one of the look-out men suddenly observed broken water or foam ahead and on the starboard bow of the ship, which was promptly reported to the officer in charge of the bridge. The master immediately rang the telegraph bell to stop and reverse the engines, and the evidence of the engineer shows that this order was executed. The helm was also put hard over to starboard, and the ship's head went off to S.W. by W., and a portion of the headway was taken off the ship. She, however, still forged ahead, and struck upon the rocks near the north-east extremity of Belle Isle, and became fixed, soon careening over to port at an angle of 22 degrees.

That water began to make its way rapidly into the various compartments of the ship from some holes made by the rocks and straining, and the master consequently stopped the engines for fear the vessel would slip off into deep water and founder. The fog whistle was then blowing rapidly to call all the crew and passengers on deck.

That immediately afterwards the said master ordered the boats to be launched from the port side, and a number of women and children were put into them, and those in- charge of the boats were ordered to pull out to seaward so as to keep clear of the ship. One of the boats was, unfortunately, damaged alongside the ship, and about 10 of the passengers drowned. The master, officers, and seamen appear; from the evidence, to have used every effort to save and shelter the passengers, who were afterwards landed on the rocks, abreast of the ship.

With reference to the navigation of the vessel, the Court is of opinion that the master is to blame for his negligence in not using the lead when approaching the coast, and he was not justified in attempting to make for the entrance of the Strait of Belle Isle in thick weather.

The Court, after due deliberation, has come to the conclusion that the cause of the disaster must be principally attributed to the total neglect of the use of the lead, and to the master placing too much reliance on the estimated position at noon of the 21st September, when no meridian altitude was taken.

The master, Henry Truscott Skrimshire, is, therefore, adjudged to be in default, and in pursuance of the powers vested in it by the provisions of the Wrecks and Salvage Act, the Court suspends his certificate for nine calendar months from the date of the commencement of this inquiry, and the certificate of the said master is hereby suspended accordingly.

W. H. Smith, R.N.R.,

Commissioner.

We fully concur in the foregoing report and decision.

Bloomfield Douglas, R.N.R.,

Naval Assistant, Marine and Fisheries, Nautical Assessor.

M. P. McElhinney,

Nautical Adviser and Nautical Assessor.

Given in open Court, at Quebec, in the Province of Quebec, Canada, this 27th day of October 1899.

Annex to Report - ss. "Scotsman."

Formal investigation into the cause of the stranding and total loss of the British steamship "Scotsman," on the Northeast part of the Island known as Belle Isle, between Newfoundland and the Labrador coast, on the morning of the 22nd of September 1899.

This investigation was held in the Admiralty Court, Quebec, on the 23rd, 24th, 25th, and 26th days of October 1899, when Mr. F. H. Market, of Smith, Markey and Montgomery, Barristers, Montreal, appeared on behalf of the master; and Mr. W. Cook, Q.C., Mr. A. A. Cook, Q.C., and Mr. Archibald Laurie, Barristers, for the shipping company.

Witnesses from the ship were produced by the captain and examined; but none of the passengers came forward to give evidence.

The Commissioner read over the authority for holding the investigation, and, having duly subpoenaed the severed witnesses, and presented the certificated officers with statements of the case upon which the enquiry was ordered, proceeded to take evidence.

"Scotsman" (s.s.)

The "Scotsman" was a steel twin screw British steamship, built at Belfast in 1895, by Harland and Wolff, and registered in Liverpool, Great Britain.

She was the property of the British and North Atlantic Steamship Company (the Dominion Line), Messrs. Richards, Mills, and Co., of Liverpool, Great Britain, being the managing owners, and David Torrance and Co., of Montreal, the agents. Official number, 105,283.

Her dimensions were as follows: —Length, 470.7 ft.; breadth, 49.2 ft.; and depth, 31.9 ft.

The vessel was fitted with triple expansion engines of 499 horsepower, her cylinders being 22, 36 1/2, and 60, with 45 stroke, and multitubular boilers, having 18 furnaces, and a steam pressure of 180 lbs., making 74 revolutions of the engines per minute at full speed, the daily consumption of coal being 75 tons.

The master, H. T. Skrimshire held an Imperial certificate of competency as ordinary master, issued by the Lords of the Committee of Privy Council for Trade, in August 1892.

There were four officers serving on hoard, and five engineers, all with appropriate certificates, and a crew consisting of seamen, firemen and stewards, amounting to 120 hands all told. The vessel also carried a duly qualified surgeon.

From the evidence adduced at this investigation, it appears that the ship was supplied with a proper number of compasses, in good order, the standard or navigating compass being of Lord Kelvin's pattern and situated upon the upper bridge in such a position that the bearing of object's could be observed in every direction. It had been inspected before leaving port.

The ship was swung for adjustment of the compass in 1896, and a table of deviations supplied; but it was stated that the master and officers took frequent observations of the sun and stars to ascertain the changes, and these were first recorded in a scrap book and afterwards inserted in a record book by the master, for future reference. Both these books were handed into Court (Exhibit "A") and proved that the master and officers had been very careful in keeping them.

The evidence shows that 12 wooden boats were on hoard in good condition, and several collapsible boats on deck ready for use at short notice, but these were not required for service after the accident, as some of the passengers and crew landed on the rocks from the bow of the ship.

They had enough anchors and chains for the requirements of the vessel, and a proper supply of lifebelts and lifebuoys for the use of those on board.

The managers of the Company supplied the ship with all the necessary charts, and those which were in use, both previous to and at the time of the casualty, were produced in Court, and had been largely corrected up to the year 1887 upon the local chart, and for 1895 on the North Atlantic Track Chart.

The master informed the Court the Company were very liberal in the supply of charts, books, and nautical instruments.

The vessel therefore appears to have been properly and suitably equipped for the intended voyage, with everything requisite for safe navigation, and had been inspected by the Board of Trade in June, 1899, and by the Emigration officers at Liverpool, previous to leaving upon this last voyage, when several boats were swung out and lowered in the water as the ship was lying in the river at anchor.

The "Scotsman" left Liverpool on the 14th of September 1899, bound for Montreal, having on board 86 saloon passengers, 106 intermediate, and 87 steerage passengers, and a cargo consisting of general merchandise of about 2,000 tons weight and measurement, and drawing 22 feet on an even Keel. The tide was ebb at the time.

She passed the North of Ireland, out through the North Channel, making her courses good by compass from one point of land to another, after allowance had been made for deviation, and the tide being taken into consideration.

At 9.50 a.m. of the 15th of September, the vessel was abreast of Innistrahull Island, north coast of Ireland, three miles distant, and a departure was taken from the lighthouse at 10.30 a.m., the weather being misty.

The course was then set N. 64° W. by compass, with 10° of easterly deviation, to pass on the Great Circle track for Belle Isle, which lies between the coasts of Newfoundland and Labrador. Nothing particular occurred during the passage across the Atlantic, and the usual weather of that season was encountered.

Observations were obtained for latitude and longitude on

several occasions, and the position of the ship entered in the logbook day by day, either by dead reckoning or observation, and the hook was kept very well by the first officer.

The dead reckoning was worked up each day by the junior officers, and the master stated that it compared fairly well with the observations, although upon one or two days there was a discrepancy between the dead reckoning worked by himself and that ascertained by the junior officers. On the 20th of September the difference was 18 miles between latitude by observation and dead reckoning, but the master produced a paper upon which he had worked up the dead reckoning himself, which reduced this difference to 11 miles.

On the 21st of September, at noon, the position of the ship, given by the master in evidence, was in latitude 53° 27' N., and longitude 50° 04' W., but the latitude entered in the logbook was 53° 24' N.

This was brought down from the course and distance run to noon, from 10.30 a.m. of that date, when double altitudes of the sun were worked up from two morning observations, and the distance from Belle Isle at noon was estimated to be 216 miles.

No meridian altitude was taken for latitude, as the sky was overcast.

The course of the vessel was then set N. 87° W. by compass (S. 65° W., true) and the speed was estimated to be about 13.5 knots per hour.

An azimuth of the sun was obtained at 2.30 or 2.40 p.m., and the error of the standard compass ascertained to be 28 1/2° West. No observations for longitude were taken at that time. At 6.45 p.m. the course was altered to N. 89° W. by compass,

(S. 63° W. true), and at 9 p.m. the master hauled the ship five degrees to the southward, for the purpose of passing eight miles off the lighthouse on Belle Isle.

The compass course was then S. 86° W. (S. 59° W. true), which course was maintained up to the time the breakers were sighted at 2.25 a.m., of the 22nd of September. At midnight the vessel was going at 13.2 knots, the weather being overcast, and the distance from Belle Isle Light estimated to be 52 miles.

The evidence shows that the master was frequently upon deck during the night, and went to his cabin shortly before midnight, leaving orders with the chief officer that he should be called at two o'clock in the morning if the weather came on thick, or before if necessary.

A night order book was kept on board and was produced in Court. Shortly before two a.m. the master went on deck and remained there until the vessel was stranded, personally conducting the navigation of the ship.

She was then proceeding upon the same course, S. 86° W. by compass, the weather being dark, with overcast sky and wind from S.W., with a rough, uneven sea. No reduction in the speed had been made. The master states it was so dark that he could not call the horizon a clear one, but he supposed he might have seen a light 14 or 15 miles off, if one had been there, as he could see all round.

By that time, however, Belle Isle Light must have been less than 11 miles distant, but. as the steamer was proceeding in a direction obliquely to the coast, it could not have been seen in consequence of the high. land to the eastward shutting it in.

No soundings were obtained at any time, the master stating that he considered his position placed him too far off the land to necessitate soundings being taken, and that he had no idea he could get bottom in any way, and feeling perfectly confident of the position of the ship, he omitted to do so.

It is stated in the Book of Sailing Directions, that there is a bank of soundings lying E.S.E. from the North point of Belle Isle, distant 28 miles, which affords a valuable means of ascertaining the position of a vessel in foggy weather when approaching the island from the eastward.

The least water as yet found upon this bank is 86 fathoms, mud, with 105 to 125 fathoms, sand, between it and the bank which stretches to the eastward of Belle Isle.

This bank has been found to be a good guide to many. Soon after two o'clock a.m. of the 22nd, some drifting haze came over the masthead light, and at ten minutes past two, the telegraph bell was rung upon the "stand by," and a few seconds after, orders were given to the engineer on watch to reduce the speed to half, as it set in foggy.

The engines were previously making 74 revolutions a minute, which were brought down to 37 or 39, giving a speed variously estimated at from 6 to 7 1/2 knots through the water. This was principally obtained by the revolutions of the engines, as it does not appear the log was hoven for that purpose.

Two men were stationed upon the look-out, the master and second officer being upon the bridge.

A dense fog then suddenly came on, which prevented objects from being seen at any distance from the ship.

The second officer says, "The fog came on very suddenly, and remained until we struck. We could see about a skip's length off." Owens, the look-out man, says, "I think I could have seen about the length of the ship off when we ran into the fog."

At 2.25 a.m., the look-out man reported something "ahead and on the starboard," but the master did not catch all he said or what was ahead, but the second officer, who was near the master, said he thought breakers were reported, and the look-out man states that as soon as he noticed foam, he called to the man in the crow's nest, who also saw the breakers and reported to the second mate upon the bridge.

It does not appear that the high cliffs of the land were seen at the time, but the master took the breakers to be the white line at the base of an iceberg.

Orders were immediately given by telegraph to stop and reverse the engines and put the helm hard-a-star- board, and the ship's head payed off to S.W. by W., and in about from $1_{1/2}$ to $2_{1/2}$ minutes afterwards she struck.

According to the evidence of the second mate, about two or three minutes elapsed from the time the man reported breakers until the vessel struck, and the ship was moving at the rate of about $1_{1/2}$ knots per hour, the speed having been taken off by reversing the engines.

There was not much shock at the time, but a hard grating on the rocks, and the vessel at once careened over to port about 23° and remained fast on rocks. It was not known where she had gone on shore.

When it was found that the ship was stranded, the master ordered the engines to be stopped. This was a wise precaution to have taken.

Had he moved the engines again, the ship might have slid off the ledge, and sunk in deep water, which was stated to be a short distance astern of her, as in about four or five minutes after she struck, the water began to rush into the ship on the port side in great volumes through the large holes which were made in her by the rocks penetrating her side, and in about 10 minutes after, the water came in to such an extent that the engineers had to clear out of the engine-room, and could not afterwards get to the platform, and the fires were drowned out.

Every precaution appears to have been taken by the engineers before leaving to prevent an explosion.

Subsequent events proved the wisdom of the master's stopping the engines at the time, as some days after, the ship did slide off the rocks and disappear beneath the surface of the water.

Had this occurred shortly after she stranded, no doubt the loss of life would have been appalling, the night being very dark with a thick fog surrounding everything.

Immediately after the casualty, the master blew alternate blasts upon the whistle and siren to arouse the crew and passengers. Orders were given to the carpenter to sound the pumps, but the chief engineer soon reported that all the holds were filling with water rapidly.

The crew then, having come on deck, were at once employed lowering the boats in the water on the port side, as those upon the starboard side could not be used on account of the heavy list the vessel had.

The master states, in his deposition, that there was no panic, and no attempt on the part of the men to get into the boats which were all lowered and away within the space of half an hour. The women and children were placed in these boats.

Nos. 4 and 6 boats got safely clear of the ship's side and were rowed out to seaward as there was no prospect at that time of finding a landing place.

The boats afterwards returned to the ship, and the people were taken on board and landed, with all the others, by a ladder placed on the rocks from the how of the ship.

After No. 8 boat was afloat, she became damaged, either by collision with one of the other boats, or by striking against the ship's side, or by some other cause unknown, and she commenced taking in water rapidly, and the crew asked for a bailer.

Evidence shows that Edward Dobbin, quartermaster, put in the ping in the bottom of the boat. He says, "we lowered her down a little below the rail, and I jumped into her, and put in the plug. I could not tell, when she was first lowered, whether the water was coming in or not. The water came in at the bow pretty rapidly, and I sung out to the 2nd officer not to lower any more passengers. I felt the other boat strike her pretty hard on the bow."

"After that she was swamped by the water coming in so much that she settled down, and a sea came in over the stern, and she then went down quickly."

"We were all thrown into the water. The painter was fast to the bow. I got up by the lifeline."

"The passengers did not cry out much. One person I am told was hauled up the ship's side."

"Some of the boat's crew went up by the lifeline. None of the crew drowned. About all of the passengers were drowned and four saved."

"We lowered down No. 10 boat as soon as possible, so as to

save the people. Four of those saved had life jackets on. It was very dark then."

"I tried all I could to save the people. I could not say if any lifebuoys were thrown to them."

The second officer says "the first boat that I lowered, No. 8, was swamped. It was kept on the saloon deck in chocks and griped down and covered. We had no trouble in getting it out. It was a boat capable of carrying from 45 to 48 passengers."

"We lowered her level with the saloon deck and put the men in who were to take charge of her, and then put in the women and children."

"Eighteen or 20 passengers were put in that boat and four of a crew. While she was fast alongside the painter, the quartermaster said that water was coming in and gaining on them, and an extra heavy sea filled her."

"The quartermaster thought one of the other boats had damaged her."

"I sung out to one of the other boats to drop down and try to rescue the passengers. " The people were very quiet and did not shout out much. They all had lifebelts on."

"It was so dark we could hardly see the boats from the deck. All that were saved were saved by the other boats. This was in about half an hour after the ship went ashore."

"Then No. 10, the aftermost boat of all, was put out to rescue the drowning people." The master states that he did not hear of the boat being swamped until all the other boats were lowered and away.

It may appear to be unaccountable that further effort should not have been made to save more of the women who were in

No. 8 boat.

There were two boats in the water alongside the ship ahead of her, and it is said that some of the drowning persons were rescued by their crews, but ropes and lifebuoys might have been thrown to them from the ship's deck until the boats came to their assistance.

It is, however, possible that a large number of people rushed to the rail of the ship, looking over the side, and prevented the officers and seamen from giving that aid which was promptly required.

With regard to the navigation of the ship, it can be seen by the compass record book, Exhibit "A," that care was taken to ascertain the deviation of the compass on every available opportunity, that the senior officers, who held master's certificates, took independent observations for latitude and longitude, and the junior officers worked up the dead reckoning daily.

The master, however, states he did not accept the position by account, given him by the junior officer on the 21st, as he found the ship had maintained the Great Circle course.

It was recorded in the logbook and should on that account be accepted as official.

As no observation was obtained at noon of the 21st for latitude, it was necessary for the safe navigation of the vessel that the master should have taken every known precaution afterwards, when steaming towards the coast; and an officer in command of a passenger steamer cannot expect to be exonerated from blame when he neglects any responsibility.

The "Scotsman" was going at full speed, of about 13.2 knots an hour, from noon of the 21st until 2.10 a.m. of the 22nd.

It may therefore be inferred that at that time, when the vessel's speed was reduced, she could not have been more than 3½ miles from the land, as after suddenly entering the dense fog, she struck the rocks at 2.25, and it was impossible for those on board to see ahead or in any direction more than the ship's length.

It did not occur to the master that under the circumstances in which the ship was placed when the fog came on icebergs might have been encountered at any time.

The fact of the ship going into dense fog so suddenly might have indicated the proximity of land. Men experienced in the North Atlantic trade find that fog will sometimes lie like a mantle enveloping islands or portions of the mainland.

It was therefore imperative for him to have stopped the engines and taken a cast of the lead. It would also have been prudent and proper if the ship had been turned round with her head to the eastward, and steamed slowly off in that direction until daylight came in. It was courting disaster to attempt to enter the Straits of Belle Isle tinder the conditions then existing, and to the lack of taking these necessary precautions the cause of the casualty must be attributed.

The number of ships which have been stranded in the Straits of Belle Isle and Gulf of St. Lawrence during the last few years, and the loss of their valuable cargoes, has caused the insurance companies to increase the rates.

In nearly all the cases of these wrecks the overconfidence of the masters in their positions has largely contributed to the disasters, but above all other causes, it appears that the neglect of the use of the lead and running at too high a rate of speed in foggy weather, have been instrumental in the destruction of so many fine ships.

Evidence shows that after the people were landed, the master, officers, and seamen did everything they possibly could to assist in taking care of and making the passengers comfortable under trying circumstances.

The precipitous cliffs near where the ship was stranded were difficult to climb, and the women and children were got up with great difficulty, some of them having to he carried a portion of the way.

Tarpaulins and other coverings were provided as soon as possible, and stores were landed in abundance, with blankets, &c., for warmth. The sailors and stewards appear to have worked cheerfully and willingly with the officers in saving food.

On the 23rd the people from the lighthouse, which was a distance of eight miles away, came to their assistance, and on the 24th, Mr. Coulton, the keeper, sent a large number of the people to the lighthouse, the master sending provisions with them.

About noon of that day the steamer "Monterey" stopped off the lighthouse and sent a boat on shore and took five saloon passengers and one steerage on board, and the master took the opportunity of despatching all firemen and trimmers, and four stowaways, who were taken in that vessel, to England.

It appears that the day after the ship was wrecked a gang of men, supposed to be firemen, went on board, and broke into the spirit stores, and abstracted a quantity of liquors, which they secreted among the rooks,

As soon as this was discovered, an engineer, well armed, was sent on board to prevent further robbery and mischief, and the chief officer was ordered to destroy all the spirits he could find in the storeroom.

These depredations were not known to the master at the time, as he was on shore, and they were beyond his control. It appears from the diary of the master that everything that was possible was done for the comfort of the passengers after they reached the lighthouse.

Warm clothing was supplied, hot meals, and a hearty welcome from Mr. Coulton. The master expresses his gratitude to Mr. Coulton and family in the following words "These good people at the lighthouse, from the day of the wreck up to this time, have been untiring in their kindness and attention, and constant sympathy and help to all our suffering passengers and crew. Their advice, and. assistance and guidance of people across the island, has been invaluable, and has without doubt in the largest degree contributed to the safe deliverance of all from a very dangerous position here. I am afraid to think of what might have happened to many of our weaker ones were it not for their unwonted help, co-operation, and watchfulness in the matter of intercepting and inducing the several vessels to come to our assistance, to say nothing of the labour willingly endured to bring people in from the wreck, entertaining them with the best in the house, and finally assisting them to the landing jetty a distance of a mile, and seeing them all safely in the boats. May God reward them."

With reference to the report that the master of the "Scotsman" was misled by the change of the fog signal on Belle Isle, it may be said that notice of the change was published on the 27th of July, in a "Notice to Mariners," stating that a siren would be substituted for the bombs hitherto used at that station, and he should have had this information.

He states in evidence, "I did not hear any other sound except the breakers before the ship struck, neither did I hear the fog alarm on Belle Isle after we struck. It was 8 miles distant. The change of the fog alarm from a bomb to a siren did not have any influence on my navigation of the ship."

Captain Skrimshire has always borne a high character upon previous voyages, and Mr. David Torrance, the agent of the Dominion Line of steamers, stated to the Court that he had been in the employ of the company for many years, and his reputation is first class, and one of the underwriters told him not long ago that owing to the fine condition of the vessel and the reputation of the master, he was taking larger risks than usual.

In consideration of these facts, and upon the application of the master, whose conduct after the accident was praiseworthy, the Commissioner recommends that a mate's certificate should be granted to the master daring the time of suspension of his master's certificate.

W. H. Smith, R.N.R., Commissioner.

Reference - Wrecks and Salvage Act Report and Court Decision, 1899/12/08, London, Board of Trade

ROMANCE OF THE WRECK

ROMANCE OF THE WRECK

TWO SCOTSMAN PASSENGERS TO MARRY.

A German Saved the Life of a Young Lady Twice and now she is to Wed her Saviour.

(Special to The Journal.)

Hespeler, Ont., Oct. 28.—The wrecking of the Scotsman on the coast of Belle Isle on the morning of Sept. 22nd, was the means of creating a romance. Among the passengers on the steamer were Henry Swartz, a German, bound for the United States, and Edith Davis, a young English lady enroute for Hespeler, Ont. During the terrible scenes which followed the wrecking of the boat, Swartz saved Miss Davis' life twice at the risk of losing his own. Instead of going on to the States, the gallant German accompanied Miss Davis to Hespeler, and to-day at four o'clock the life which he twice saved, will by the holy bonds of wedlock be united to his until death doth them part.

ADDITIONAL BOOKS BY PAUL LANGAN

All books are available at Wordsworth Books, Waterloo Ontario, from the Author through his website, http://paullangan.com, and Amazon worldwide, Rakuten/Kobo worldwide.

<u>Forgotten Monorails of Canada</u> - Released Dec. 6th, 2020, ISBN - 978-0968586761

The definitive history of monorails in Canada from 1896 to 2020. From Vancouver BC to Cavendish PEI, Canada has had 6 monorails that have been in operation in the past. Two monorails are still in operation. What makes this book unique is it also reviews all the monorail proposals that have been pitched in over the years and there have been a lot of them. Vancouver, Edmonton, Calgary, Winnipeg, Niagara Falls, Toronto, Montreal are some of the cities that had studies done looking at monorail as a possible transit system. Montreal alone has had over 15 studies that included monorails as a possible option. Did you know that demonstration lines were built in Fort Saskatchewan and Niagara Falls? Low-speed and High-speed monorails ideas have been pitched. You won't find this topic covered on Wikipedia or in a Google search.

<u>Tragedy in Galt - The May 2nd, 1956, CP Rail Crash</u> - 2nd edition released, Jan. 24th, 2021 ISBN - 978-0968586778

The definitive book on the tragic CP Rail Crash May 2nd, 1956, in Galt, Ontario (now Cambridge). William Palmer and Tommy Watson lost their lives in the crash. Interviews with 4 members of the train crews that were there that day. Sixty-four years later people are still haunted by the images of that fateful day. Includes the full Inquest, photos and interviews with Ken Shillington, William Hopkins, Ron Westworth and George Nutkins.

<u>Remembering Glen Christie: The history of the Village of Glen Christie, Ontario and stories from those that lived there</u> - 2nd edition released, April 6th, 2022,

ISBN - 978-0968586785

Remembering Glen Christie is about a now vanished village that was north of Hespeler (Cambridge), Ontario. The history, photos, stories and memories of those that lived there. With guest contribution from Puslinch Township historian Marjorie Clark.

<u>Tales of Opera - Waterloo County/Region 1885-2020</u> - Released August 19th, 2022,

ISBN - 978-1777864309

The history of opera and operettas in the Waterloo County/Region in Ontario Canada from 1880-2020.

Features chapters on: Berlin Opera House, Scott's Opera House, Twin City Operatic Society, Mabel Krug, Preston Operatic Society, Gilbert & Sullivan Society, Opera at Wilfrid Laurier University, Kitchener-Waterloo Opera Guild, Kitchener-Waterloo Oktoberfest Operettas, Opera Ontario, K-W Opera, Kitchener Opera, Vera Causa Opera and so much more.

Guest writers - Leslie De'Ath and Dr. Ted Rhodes.

Hespeler History - The Authorized Winfield Brewster Collection - Released October 20th, 2021, ISBN 978-1777864316

The definitive history on the early history of Bergeytown, New Hope, and Hespeler, Ontario by Winfield Brewster. Featuring the following booklets:

- J. Hespeler, New Hope C.W. - 1951
- The Floodgate: Random Writings of Our Ain Folk - 1952
- Hespeler Yarns - 1953
- La Rue de Commerce; Queen St. Hespeler, Ontario, - 1954
- Plus The Short History of Hespeler Public School and rare Maps and Photos

Classic Hockey Stories - From the Golden Era of Pulp Magazines 1930s-1950s -Released November 10th, 2021, ISBN 978-1777864323

Including: Blazing Blades - Barry Kevin, Blonde Bullet (novelette) - Giles A Lutz, Charge of the Ice Brigade (rink novel) - Joe Archibald, Double-Backed Puckster - Ralph Powers, Hockey Horoscope 1938 NHL Season - Jack Kofoed, Pardon My Puck - T W Ford, Pucksters on the Prod - Mac Davis, Blue Line Blazers by Theodore J. Roemer, Stooge for Puck Pirate (novelette) - by C. Paul Jackson, Maurice "The Rocket" Richard Hockey's Battling Terror Comic, Eric Laprade, Gentleman of the Rink - Comic, Top Hockey Stars 1950 - Comic, Authors, Canadiana Further Reading.

Sabotage in Hespeler - (a short story) -Released 12/04/2021, ISBN 978-1-7781289-4-3

FEATURING- WWI was raging and our Canadian boys were fighting and dying overseas. On February 3rd, 1916, our Parliament building in Ottawa burned to the ground under suspicious circumstances. On

February 6th, 1916, the A.B. Jardine Munitions Factory in Hespeler, Ontario was destroyed by fire. Was the fire set by German spies? Paul Langan weaves a creative non-fiction short story based on the facts uncovered that fateful day in Hespeler. (3000 words)

Alternatives Denied - The $100,000,000 Highway 7 Fiasco (2000) - REPRINT

Released, March 28th, 2022, ISBN 978-1-7778643-6-1.

The provincial government announced recently they would move forward with the plan to build a new Highway #7 between Kitchener and Guelph. A group of concerned citizens put together a book in 2000 about the fiasco. The project was stopped for two decades. The reprint of this book is a chilling reminder of a flawed rationale to build a new highway.

Play and Coach Hockey Like 1959! - Release date, August 24th,2022, ISBN: 978-1-7781289-5-0

FEATURING: In 1959, The Royal Canadian Air Force released a series of booklets on hockey titled, Beginning Hockey, How to Play Better Hockey and Coach's Manual. They included skill development, instruction and drills for the player and coach. These booklets are being reprinted in their entirety for the first time in half a century. Go "old school" and read this book. Learn how hockey was played and taught in 1959. - Paul Langan

The History of Fisher Mills and Beaverdale, Ontario - Release date, Oct.2nd, 2022, ISBN 978-1-7778643-4-7.

FEATURING: History of Fisher Mills, the mill, dam and pond, Edward Snyder & Snyder's Potato Chips, The Hotels, Stager General Store and Blacksmith, Maple Grove School, Zion United and Wanner Mennonite Church, Squire Ellis, The Beaver Family, Ed Ratcliffe Historic homes still standing, rare photos and maps and more!

VIA Rail Canada vs The Public - Release date, Nov.2nd, 2022, ISBN:978-1778128981

FEATURING: VIA Rail Canada vs the Public: VIA Rail denied Canadians the right to buy the CANRAILPASS. A look back at this 1991 debacle. 116 pages, in 1991, a citizen was refused a VIA RAIL Canada "CANRAILPASS" because they were a Canadian citizen. Paul Langan and MP Herb Gray filed a complaint with the National Transportation Agency under the Railway Act that Canadians were being prejudiced against by VIA Rail Canada for prohibiting them from buying the pass.

Absurd as this sounds, VIA Rail Canada fought against offering the pass to Canadians and even went as far as fighting it, and losing, at the Federal Court of Appeal. Featured in the Book:

The National Transportation Agency Decision, The National Transportation Agency Order,

The National Transportation Agency Staff Report, The Federal Court of Appeal Decision and

Newspaper Articles

Made in the USA
Monee, IL
04 November 2026

67375698R00039